# What's So Funny about Science?

Cartoons by
**Sidney Harris**

From
**American Scientist**

Foreword by
**Herbert S. Bailey, Jr.**

**WILLIAM KAUFMANN, Inc.**
Los Altos, California

# *Dedication*

Special thanks to Jane Olson, Editor of *American Scientist*, who made this collection possible.

Library of Congress Catalog Card No.: 77-82638

Harris, Sidney

What's so funny about science?

Los Altos, Ca.: Wm. Kaufmann, Inc.

120 p.

7707          770621

ISBN 0-913232-39-4     ISBN 0-913232-42-4 (pbk)

Printed in the United States of America

# *Foreword*

Pollution is a serious problem. We are filling our oceans with oil and sewage and mercury and junk. Then why is it funny when one dolphin says to another, "There was a time when I thought humans were as smart as we are"? We know that dolphins have a language; maybe that's what they're really saying.

And why is it funny when one angel says to another, "I'm beginning to understand eternity, but infinity is still beyond me"? Is it because, even though we have a symbol for infinity and can manipulate it mathematically, we suddenly realize that *we* don't understand it either? Or do we smile because we think we do?

And why are we amused at the thought of Professor Einstein at the blackboard writing $E = ma^2$, crossing it out, then trying $E = mb^2$, crossing it out again — and we realize that the great discovery of the atomic age is about to be revealed. Science isn't like that; in fact science isn't usually funny at all.

Psychologists say that humor arises from the unexpected and the incongruous. It surprises us into a new point of view — telling us that perhaps dolphins *are* smarter than we are, or that we *don't* really understand infinity (or eternity either), or that science is a lot more than the manipulation of equations.

S. Harris likes science and thinks it is serious, but he also thinks a lot of things about science are funny; even the dreadful things like pollution or the possibility of an atomic accident have their humorous aspects, their logical (or illogical) contradictions.

One of the funny things about science is the scientists themselves. Even though we usually take scientists quite seriously, and even though they often take themselves *very* seriously, S. Harris thinks they are funny — sometimes. And by his stretches of imagination and his inspired pen he shows us *how* they are funny.

S. Harris likes animals too, especially the intelligent ones that have learned so much from the scientists who have experimented on them — the rats in a maze, the monkeys who do arithmetic to get bananas, the jungle animals who talk to each other. In fact all of Harris's animals talk to each other, giving us a point of view on science that we might not have thought about.

But what's so funny about science? If you want to know, look at S. Harris's cartoons, which originally appeared in the *American Scientist* magazine along with articles like "The Origin and Influence of Airborne Particulates." With Harris's help, scientists are able to laugh at themselves and at science (or at least Harris's version of it). Three or four of Harris' cartoons have appeared in each issue of the *American Scientist* since January, 1970; like all humor they are best taken in small doses. It's a scientific fact that with repeated stimulation the funny bone gets numb. My advice to scientists is to keep this book on your desk or in your laboratory, and when that proof doesn't work out or the experiment goes wrong, pick it up and look at it for a few minutes. It won't solve your problem but it will make you feel better.

*Herbert S. Bailey, Jr.*

Herbert S. Bailey, Jr., Director of the Princeton University Press, is a member of the Publications Committee of Sigma Xi, the Scientific Research Society of North America, publishers of *American Scientist*.

# About the Artist

Sidney Harris was born in the good old days. He came to cartooning via P.S. 177 in Brooklyn (where he was held captive in World War II) and the Art Students League in New York. Except for a summer job as a messenger, which lasted three weeks, and five nights as an extra (supernumerary) at the Metropolitan Opera, he has avoided employment. Even though he is duly impressed by the drawings of Rembrandt and Hokusai, to name but two, this has not deterred him from free-lancing, and he produces his own kind of art for *Playboy*, *The New Yorker*, and for various other periodicals in addition to *American Scientist*. He generally draws away in his attic in Great Neck, while his wife, Marilyn, does her sculpture and ceramics in her basement and their children, Jonathan and Jennifer, run up and down. His previous cartoon collections are: *So Far, So Good* (Playboy Press, 1971) and *Pardon Me, Miss* (Dell Publishing Co., 1973).

"The big bang? Believe me, it was very, very, very, very, very, *very* big."

"These days *everything* is higher."

"What do you expect, since 90% of all the scientists
who ever lived are alive today?"

"But technology has created an information explosion —
everyone *does* have to talk at once."

"Let's face it. Evolution has passed us by."

"Gottlieb, I think I know why we've been receiving so
few commissions."

"I have a feeling it's too soon for fossil fuels around here."

"What's the big deal about solar heat?"

"Our problem, once solar energy is in operation, is to find a way to have the citizens whose homes are heated by the sun continue to pay *us* every month."

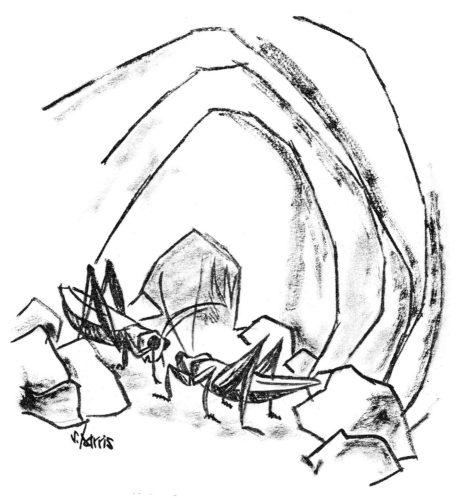

"I can't *stand* this waiting. Couldn't we be *six*-year locusts?"

"The report isn't all bad. They say that, eaten with warm milk, the *box* could provide some nutrition."

"With the new, vitamin-enriched formula, you'll get an
additional five miles per feed bag."

"We have reason to believe that Bingleman is
an irrational number himself."

"What I especially like about this baby is this little
drawer where I can keep my lunch."

"You did very well on your I.Q. test. You're a man of 49
with the intelligence of a man of 53."

"Frankly, I don't see how we can keep it burning
through eternity."

"This is, without a doubt, the earliest piece of pottery found on the North American continent which was used for ornamental purposes and was made in Hong Kong."

"But Gershon, you can't call it Gershon's equation if everyone has known it for ages."

"Actually I started out in quantum mechanics, but
somewhere along the way I took a wrong turn."

"AH, HA!"

"Take it from me and come back. The future
is definitely on land."

"Grantz is charting his life based on genetic
vs. environmental factors."

"I don't care what it looks like — they're pulsars."

"Life, yes — but as for intelligent life, I have my doubts."

"Some of these youngsters have come up with a terrific
new idea — feathers."

"It may very well bring about immortality,
but it will take forever to test it."

"Dr. Chambers is unscrambling messages from outer space, Dr. Waddell is working on computer language, and Dr. Saville has been conversing with dolphins, but perhaps you could all find some common form of *human* communication."

"Now, with the *new* math . . ."

"... and in 1/10,000 of a second, it can compound
the programmer's error 87,500 times!"

Multi Independent Retarget Vehicle
Multiple Independent Reentry Vehicle
Multiple Independently Reentering Vehicle
Multi Inter-directional Reentry Vehicle
Multi Inter-target Reentry Vehicle
Multiple Inter-target Retarget Vehicle
Multi Independent Reentry Vehicle
Multiple Independently-targetable Reentry Vehicle
Multi Independent-target Reentering Vehicle
Multiple Independent Retargetable Vehicle
Multi Inter-target Reentry Vehicle
Multiple-target Inter Reentry Vehicle
Multiple Inter-Retargetable Vehicle
Multi Independent Retarget Vehicle
MIRV
S. harris

Faster than a speeding bullet:
ds
-0 s → +
More powerful than a locomotive:
HP = 33,000 ft lb / min
Able to leap tall buildings in a single bound:
KE = ½ mv²
S. Harris

"I think you should be more explicit here in step two."

"Just between you and me, where does it get enriched?"

Chapter 7.  THE STRUCTURE OF THE NUCLEUS OF THE ATOM
   "What?" exclaimed Roger, as Karen rolled over on the bed and rested her warm body against his.  "I know some nuclei are spherical and some are ellipsoidal, but where did you find out that some fluctuate in between?"
   Karen pursed her lips.  "They've been observed with a short-wavelength probe . . ."

"Oh, for Pete's sake, let's just get some ozone
and send it back up there!"

"As I understand it, they want an immediate answer. Only trouble is the message was sent out 3 million years ago."

"Due to a tightening of the budget, we are forced to curtail our overtime and weekend schedule, and request that all major breakthroughs be achieved as early in the week as possible."

"I tend to agree with you—especially since $6 \cdot 10^{-9} \sqrt{t_c}$ is my lucky number."

"Frankly, I'd be satisfied now if I could even
turn gold into lead."

"But we just don't have the technology to carry it out."

"I thought continental drift was much slower."

"It was bound to happen.—
they're beginning to think like binary computers."

"Treadmills! Mazes! There must be more to life than this."

"Sorry I'm late — I was working out $\pi$ to 5,000 places."

"The devil with the food chain. I *like* mercury."

"To think our field trip in Physics 1 was to a boiler factory."

"Oh, that's not Dr. Gershenzon. That's a hologram."

"We struck flounder."

"If we ever intend to take over the world, one thing we'll
*have* to do is synchronize our biological clocks."

"Today's problems should have been solved in the 1950s, but in the 50s we were solving the problems of the 20s, in the 20s we were solving the problems of the 1890s . . ."

"The codeine is O.K., and the phenobarbitol is O.K.,
but the Food and Drug Administration says no
to powdered bat's tooth."

"Discover any new prime numbers lately?"

"That wraps it up — the mass of the universe."

"The supporting beams and evidence of tracks lead to the conclusion that the caveman used this tunnel for the express line of his subway system."

"Some of us endangered species are getting together
Saturday night for one last fling at the water hole."

"Remember — it's better to light just one little thermonuclear
power station than to curse the darkness."

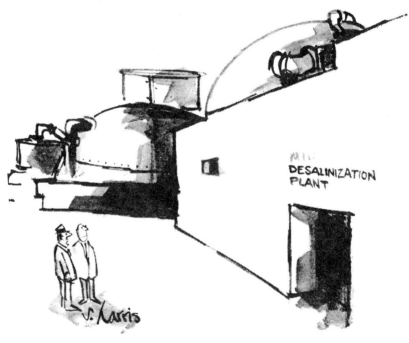

"Just in case it doesn't work, I'd like you to come up with some uses for 10 million gallons of salt water every day."

"If it's true that the world ant population is $10^{15}$, then
it's no wonder that we never run into anyone we know."

"What's the difference? It was a man-made lake
in the first place."

"Well, if *you* can't use it, do you know anyone who *can* use 3,000 tons of sludge every day?"

"Believe me, after a hard day in the jungle, it's a pleasure being shot by one of those scientists with a tranquilizer gun."

"Why don't you check with the local databank?"

"Clam chowder — Ingredients: clams, potatoes, water, hydrolated plant protein, sodium phosphate, calcium carbonate, butylated hydroxytoluene. For external use only."

"The only other solution is that we may evolve into
a species immune to all this junk."

"Go right ahead. I realize you're bred to be violent, just
as I'm bred to be passive."

"Look," I would say to Leonardo. "See how far our technology has taken us." Leonardo would answer, "You must explain to me how everything works." At that point, my fantasy ends.

"Perhaps, Dr. Pavlov, he could be taught to seal envelopes."

"All these delays — a thousandth of a second here, a millionth of a second there — we'll have to get the darn thing fixed."

"Now try to find a few that look like a bear
or a dog or something."

"The generic name for Meplosutricin? Snake-skin oil."

"Matthew want eat. Get banana. Get bread. Get milk."

"Somehow I was hoping genetic engineering would take
a different turn."

"Vacuums, black holes, antimatter — it's the elusive and
intangible which appeals to me."

"Leopards! Zebras! But what if *rhododendrons* became extinct?"

"From all indications, ours is one of the biggest
universes there is."

"What's most depressing is the realization that everything
we believe will be disproved in a few years."

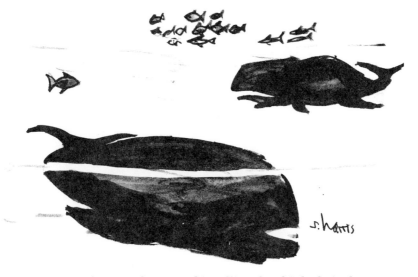

"Just because his record is selling, he thinks he's the
only one who knows how to sing."

"This one writes some fine lyrics, and the other one has done some beautiful music, but they just don't seem to hit it off as collaborators."

"Actually, they all look alike to me."

"This place is all right. Two more weeks, and I'll
be a molecular biochemist."

"Does this apply always, sometimes, or never?"

"There was a time I thought humans were as smart as we are."

"Take it easy, buddy — I'm not in your pecking order."

"Congratulations. They've named a lung disease after you."

"I don't use it anymore, since I got my microwave oven."

"Our sun is more than four billion years old and has already reached about half its life expectancy. It is now time to plan for the future of mankind, and a positive first step is the election of someone who is willing to face this vital problem..."

"I'm beginning to understand eternity, but infinity is still
beyond me."

"They don't even know I exist — I cause hiccups."

"I've tried it, Ned. Kicking doesn't work. There must be some *other* way to get oil out of shale."

"Although humans make sounds with their mouths and occasionally
look at each other, there is no solid evidence that they
actually communicate among themselves."

"Our plan is to extract sulphates, bromides, copper, silver, and gold from sea water. All we've managed to get so far, however, is salt."

"Look — there goes one of those U. F. O.'s again."

"As I see it, it's a toss-up between a Belgian data processing machine and an American electronic computer."

"I do hope they do get along. Fred is binary and Eric is digital."

"... and the record low for this date is 147° below zero,
which occurred 28,000 years ago during the Great Ice Age."

"You don't seem to understand, Prescott. We're not trying
to cure diseases occurring *only* in guinea pigs."

"You win a little and you lose a little. Yesterday the air didn't look as good, but it smelled better."

"To think we used to complain about a little
flotsam and jetsam."

"This must be Fibonacci's."

If left to itself a body will move with uniform velocity in one and the same direction.
GALILEO – I SAID A RIGHT AT THE CORNER, AND THEN A LEFT!
S. HARRIS

"I guess you can say that since Leeuwenhoek our family
has been in show business."

"The Periodic Table."

"That 187,000 miles per second makes me a bit skeptical
about the whole thing."

"We know he didn't discover that new virus — we're just naming it after Rheinblatt because it *looks like* Rheinblatt."

"On the other hand, my responsibility to society makes me
want to stop right here."